Tucholsky Wagner Zola Scott Sydow Freud Schlegel
Turgenev Wallace Fonatne
Twain Walther von der Vogelweide Fouqué Friedrich II. von Preußen
Weber Freiligrath Frey
Fechner Ernst Frommel
Fichte Weiße Rose von Fallersleben Kant
Richthofen
Engels Fielding Hölderlin
Fehrs Faber Flaubert Eichendorff Tacitus Dumas
Eliasberg Ebner Eschenbach
Feuerbach Maximilian I. von Habsburg Fock Eliot Zweig
Ewald Vergil
Goethe London
Mendelssohn Balzac Shakespeare Elisabeth von Österreich Ganghofer
Lichtenberg Rathenau Dostojewski
Trackl Stevenson Doyle Gjellerup
Mommsen Tolstoi Hambruch
Thoma Lenz Hanrieder Droste-Hülshoff
Dach Verne Hägele Hauff Humboldt
Karrillon Reuter Rousseau Hagen Hauptmann Gautier
Garschin
Damaschke Defoe Hebbel Baudelaire
Descartes Hegel Kussmaul Herder
Wolfram von Eschenbach Darwin Dickens Schopenhauer Rilke George
Bronner Melville Grimm Jerome Bebel Proust
Campe Horváth Aristoteles Federer
Bismarck Vigny Barlach Voltaire Herodot
Gengenbach Heine
Storm Casanova Tersteegen Grillparzer Georgy
Chamberlain Lessing Langbein Gilm
Brentano Lafontaine Gryphius
Strachwitz Claudius Schiller Kralik Iffland Sokrates
Bellamy Schilling
Katharina II. von Rußland Gerstäcker Raabe Gibbon Tschechow
Löns Hesse Hoffmann Gogol Wilde Gleim Vulpius
Luther Heym Hofmannsthal Klee Hölty Morgenstern
Roth Heyse Klopstock Kleist Goedicke
Luxemburg Puschkin Homer Mörike
La Roche Horaz Musil
Machiavelli Kierkegaard Kraft Kraus
Navarra Aurel Musset Kind Moltke
Nestroy Marie de France Lamprecht Kirchhoff Hugo
Laotse Ipsen Liebknecht
Nietzsche Nansen Ringelnatz
Marx Lassalle Gorki Klett Leibniz
von Ossietzky May Lawrence Irving
vom Stein Knigge
Petalozzi Platon Pückler Michelangelo Kock Kafka
Sachs Poe Liebermann
de Sade Praetorius Mistral Zetkin Korolenko

The publishing house tredition has created the series **TREDITION CLASSICS**. It contains classical literature works from over two thousand years. Most of these titles have been out of print and off the bookstore shelves for decades.

The book series is intended to preserve the cultural legacy and to promote the timeless works of classical literature. As a reader of a **TREDITION CLASSICS** book, the reader supports the mission to save many of the amazing works of world literature from oblivion.

The symbol of **TREDITION CLASSICS** is Johannes Gutenberg (1400 – 1468), the inventor of movable type printing.

With the series, tredition intends to make thousands of international literature classics available in printed format again – worldwide.

All books are available at book retailers worldwide in paperback and in hardcover. For more information please visit: www.tredition.com

tredition was established in 2006 by Sandra Latusseck and Soenke Schulz. Based in Hamburg, Germany, tredition offers publishing solutions to authors and publishing houses, combined with worldwide distribution of printed and digital book content. tredition is uniquely positioned to enable authors and publishing houses to create books on their own terms and without conventional manufacturing risks.

For more information please visit: www.tredition.com

Tea Leaves

Francis Leggett

Imprint

This book is part of the TREDITION CLASSICS series.

Author: Francis Leggett
Cover design: toepferschumann, Berlin (Germany)

Publisher: tredition GmbH, Hamburg (Germany)
ISBN: 978-3-8491-8502-2

www.tredition.com
www.tredition.de

Copyright:
The content of this book is sourced from the public domain.

The intention of the TREDITION CLASSICS series is to make world literature in the public domain available in printed format. Literary enthusiasts and organizations worldwide have scanned and digitally edited the original texts. tredition has subsequently formatted and redesigned the content into a modern reading layout. Therefore, we cannot guarantee the exact reproduction of the original format of a particular historic edition. Please also note that no modifications have been made to the spelling, therefore it may differ from the orthography used today.

TEA LEAVES

By
Francis Leggett & Co.

PREFATORY

The casual reader in many a nook and corner of this extended land will perhaps ask — "Who are the publishers of this book, and what is their purpose?" We anticipate any such enquiry, and reply that Francis H. Leggett & Co. are Importing and Manufacturing Grocers; that our object in publishing this and other books is to bring ourselves and our goods into closer relations with consumers at a distance from New York; and incidentally, to provide readers with interesting information respecting the food which they eat and drink.

In our search for material to aid in the preparation of this book, we were greatly indebted to Mr. F. N. Barrett, editor of THE AMERICAN GROCER, who generously gave us access to what is probably the most complete and valuable collection of books upon Foods to be found on this continent.

We wish to also to acknowledge the kind response of Messrs. Gow, Wilson and Stanton, of London, to our requests for statistics of the World's Tea Trade, and particularly for information respecting the Teas of Ceylon and India. If our limitations of space had permitted, we should have materially increased the interest of our little book by additional matter derived from the last named firm.

(Omitted) Our colored Frontispiece is a faithful representation of a Chinese tea plant, showing the flower and the seeds.

TEA LEAVES

"Pray thee, let it serve for table-talk." — Merchant of Venice.

"A cup of tea!" Is there a phrase in our language more eloquently significant of physical and mental refreshment, more expressive of remission of toil and restful relaxation, or so rich in associations with the comforts and serenity of home life, and also with unpretentious, informal, social intercourse?

If rank in the scale of importance of any material thing is to be determined by its extensive and continued influence for good, to tea must be conceded a very elevated position among those agencies which have contributed to man's happiness and well- being.

Most remarkable changes have occurred in the production of tea during the past century. About sixty years ago all the tea consumed on the globe was grown in China and Japan. Our knowledge of the growth and manufacture of tea was then of an uncertain and confused character, and no European had ever taken an active part in the production of a pound of tea. To-day, about one-half of the tea consumed in the world is grown and manufactured upon English territory, on plantations owned and superintended by Englishmen, who have thoroughly mastered every detail of the art, while nearly all the tea drank in Great Britain is English grown. Twenty years ago, the suggestion that tea might yet be grown upon a commercial scale in the United States was received with derision by the Press and its readers; but one tea estate in South Carolina has during the past year grown, manufactured, and sold at a profit, several thousand of the tea of good quality, which brought a price equal to that of foreign fine teas.

A natural taste for hot liquid foods and drinks is common to all races of men, and they may be traced in the soups of meat and fish, and in their decoctions or infusions of vegetable leaves, seeds, barks, etc.

Hot "teas" were in habitual use as beverages among civilized nations long before they ever heard of Chinese tea, of coffee, or of cocoa. The English people, for instance, freely indulged in infusions of Sage leaves, of leaves of the Wild Marjoram, the Sloe, or blackthorn, the currant, the Speedwell, and of Sassafras bark. In America, Sassafras leaves and bark were used for teas by the early colonists, as were the leaves of Gaultheria (Wintergreen), the Ledums (Labrador tea), Monarda (Horsemint, Bee-balm, or Oswego tea), Ceanothus (New Jersey tea or red-root), etc. Charles Lamb, in his essay upon Chimney Sweeps, mentions the public house of Mr. Reed, on Fleet street in London, as a place where Sassafras tea (and Salop) were still served daily to customers in his time, about 1823. Mate, Yerba, or Paraguay tea has been a national beverage for millions of people in the central portions of South America for several centuries.

With the exception of Mate, not one of the above named substitutes for Chinese tea contains the peculiar nerve stimulating and nerve refreshing constituent upon which depends the physiological value of Black or Green tea, the Theine: nor do they possess the characteristic flavoring principle or essential oil which distinguishes commercial teas from all other known plant products. The Ledums are indeed accredited by Professor James F. Johnson (Chemistry of Common Life) with stimulating and narcotic properties, but the same may be said of tobacco.

A comforting, stimulating and healthful beverage, which has been in habitual use by the most extensive nation of the globe for more than a thousand years, and which has at length become a necessity as well as a luxury for seven hundred millions of people, or of a majority of the inhabitants of the earth, is certainly worthy of more than the passing thought which accompanies its daily use in the form of "cup of tea."

Douglass Jerriold, writing of tea, some 50 years ago, said:— "Of the social influence of Tea upon the masses of the people in this country, it is not very easy to say too much. It has civilized brutish and turbulent homes, saved the drunkard from his doom, and to many a mother, who else have indeed been most wretched and forlorn, it has given cheerful, peaceful thoughts that have sustained her. Its work among us in England and elsewhere, aye, throughout the civilized world, has been humanizing and good. Its effect upon us all has been socially healthful; peaceful, gentle and hearty."

There is no article of common use about which so little is popularly known, or of which "we know so many things which are not so." The very names of the various kinds of tea which we use are mysteries of meaning to those who have not made special researches into the subject. And the cause of the distinctions in the qualities of different teas, as of black and green, are still matters of uncertainty and controversy among many dealers of teas, as well as among unscientific travelers and some untraveled scientists. The enthusiastic collector of writings upon tea by self qualified experts, will find himself involved in a maze of contradictory assertions and opinions from which there is no escape save by the exercise of judicial powers, by an independent exercise of his own judgment, in separating truth from error. And unless he is a proficient in physiology and chemistry, he will find himself baffled at last, because several important scientific questions concerning Tea are still unsolved by adequate authority.

Then there are otherwise sane persons who profess to discover in the habitual use of tea by whole nations a cause of national deterioration. We record the fact as one of the curiosities of mental perversity in an age of general intelligence.

How the inestimable qualities which lie latent in the green leaf of the Tea tree or bush were discovered and developed by the Chinese is one of those mysteries which we shall never solve. For it is a remarkable fact that neither the green leaf of the tea plant, nor the tea leaf dried without mans agency, conveys to human senses any hint of the agreeable or valuable qualities for which tea is esteemed, and which have been developed by the art of man. A leaf of any one of the mints, or of the sassafras tree, or of the wintergreen vine, after being bruised in the hand and applied to the nose or the mouth, makes instant impression upon the senses of taste and small, and at once informs us of its distinctive qualities. Not so with the tea leaf; a hundred valueless plants impress those senses more vividly than the leaf which is worth them all. Infuse the green leaf of the Tea plant and the prized properties of "Tea" are still wanting, but in their stead, positively deleterious qualities are said to appear in the infusion. Commercial Tea must be regarded as an artificial production. A certain degree of artificial heat, of manipulation, and induced chemical changes, are the agents which develop the flavor and aroma of the tea leaf. And the nature of man's treatment and manipulation determines in large measure not only the desired flavor, but the distinguishing character of the tea, its rank as a green, a black, or an "English Breakfast Tea," all three of which may be evolved by skilful manipulation from the same tea bush, at the same time.

Much has been said and written in contention upon this latter assertion, and books may be quoted upon either side of the question, but we make the statement without qualification and upon unquestionable authority.

As Chinese teas became known to the inhabitants of other parts of Asia, and to Europeans, curiosity and commercial interests impelled other races to seek information concerning the origin and treatment of different Chinese teas. The prices obtained by the Chinese from foreigners for teas two and three centuries ago were most exorbitant, and paid the Chinese Government and Chinese merchants an enormous profit. Quite naturally that sagacious nation saw the danger of letting the truth concerning the origin, manufacture and cost of their most precious commodity pass into the possession of other people, and they strove to prevent foreigners from penetrating to their inland tea gardens, while they plied inquisitive enquirers with fairy tales which were eagerly swallowed. They said that every different kind of tea was the product of a different species of plant, which bore a different name, and that the manufacture was a most intricate process depending upon secrets confined to a very few; that the leaves could safely be plucked only at certain phases of the moon, and at certain hours of the day, and that some delicate varieties of tea leaves were plucked only by young maidens, etc. They even allowed Europeans to believe that green tea was colored by salts of copper, on copper plates, having doubtless learned that their were European merchants who would not be deterred from vending poisonous foods provided a good fat profit attended the transaction. In short, they practiced some of the dissimulation and tricks of trade to which many merchants were addicted.

To particularize further, and yet generalize at the same time, we will say here that the Tea plant or tree is greatly modified in hardiness, in height, in size of leaf, and in the quality of the leaf for a beverage, by soil, by moisture, tillage, and climate. Some soils and some climates develop a tea plant decidedly more suitable for a green tea than for a black tea, and vice-versa. The Formosa Oolong, with its natural flowery fragrance is a product of a peculiar soil, said to be a clay topped with rich humus. Analysis would probably disclose peculiarities in that soil not yet found in other tea districts. In removal to other soils and other localities, the Formosa Tea plant loses its most precious characteristic, its sweet flowery aroma and taste. The total product of this tea is but 18,000,000 lbs. per annum, an insignificant quantity compared with the aggregate crops of Chinese or of Indian tea gardens. If the exceptional characteristics of Formosa Oolong accompanied the plant when removed to other localities, its cultivation would quickly become greatly extended.

What is known or believed concerning the remote history of Tea and of its dissemination among other nations than the Chinese and Japanese, has been told so often that its recapitulation becomes tedious to those who are familiar with the story. But this book is intended for the general reader, and for the purpose of collecting and welding together disconnected and floating facts and scraps of tea literature gathered from many sources.

CHAPTER II.

HISTORICAL.

Until a quite recent period botanists believed that the tea plant was a native of China, and that its growth was confined to China and Japan. But it is now definitely known that the tea plant is a native of India, where the wild plant attains a size and perfection which concealed its true character from botanical experts, as well as from ordinary observers, for many years after it had become familiar to them as a native of Indian forests.

How early in the history of the Chinese that people discovered and developed the inestimable qualities of the tea plant is not known. That Chinese scholar, S. Wells Williams, in his Middle Kingdom places the date about 350 A.D. But somewhere between 500 A.D. and 700 A.D. Tea had become a favorite beverage in Chinese families. Some of the written records of that ancient people push the epoch of tea-drinking back as far as 2700 B.C., appealing to ambiguous utterances of Confucius for corroboration. Tea in China had obtained sufficient importance in political economy in 783 or 793 A.D. to become an object of taxation by the Chinese Government.

Gibbon, in his great work, tells us that as early as the sixth century, caravans conveyed the silks and spices and sandal wood of China by land from the Chinese Sea westward to Roman markets on the Mediterranean, a distance of nearly 6,000 miles. But we hear no mention of the introduction of tea into Europe or western Asia until a thousand years later.

According to Mr. John McEwan (International Geog. Congress, Berlin, 1899,) tea soon found its way from China into Japan and Formosa, but was not cultivated in Japan on a commercial scale until the 12th century.

John Sumner, in a Treatise on Tea (Birmingham, 1863), states that the Portuguese claim to have first introduced tea into Europe, about

1557. Disraeli (Curiosities of Literature) offers evidence that tea was unknown in Russian Court circles as late as 1639.

But Russia and Persia seem to have naturalized tea as a beverage about the same time that it became known in England. Little is said about Persian tea-drinking in modern writing upon tea, but the testimony of many travelers bears witness to the national love of tea by Persians.

The Encyclopedia Britannica concedes to the Dutch, the honor of being the first European tea-drinkers, and states that early English supplies of tea were obtained from Dutch sources. It is related by Dr. Thomas Short, (A Dissertation on Tea, London, 1730), that on the second voyage of a ship of the Dutch East India Co. to China, the Dutch offered to trade Sage, as a very precious herb, then unknown to the Chinese, at the rate of three pounds of tea for one pound of Sage. The new demand for sage at one time exhausted the supply, but after a while the Orientals had a surfeit of sage-tea, and concluded that Chinese tea was quite good enough for Chinamen. If the European traders had known the virtue of sage-tea for stimulating the growth of human hair, and had given the Orientals the cue, sage leaves might have retained their high value with the Chinese until now.

In these days, it may be remarked, the Dutch are said to drink as much tea per capita as the Russians, who are as fond of tea as the Chinese.

While both the English and Dutch East India Companies exhibited in England small samples of tea as curiosities of barbarian customs very early in the 17th century, tea did not begin to be used as a beverage in England even by the Royalty until after 1650.

In a number of the weekly Mercurius Politicus (a predecessor of the present London Gazette), dated September 30, 1658, occurs this advertisement:

"That excellent and by all pysitians approved China drink called by the Chineans Tcha, by other nations Tay, alias Tee, is sold at the Sultaness Head, a Cophee-house in Sweetings Rents, by the Royal Exchange, London."

This appears to be the earliest recorded and authentic evidence of the use of tea in England.

Macaulay, in a note in his History of England, says that tea became a fashionable drink among Parisians, and went out of fashion, before it was known in London, and refers to the published correspondence of the French physician, Dr. Guy Patin, with Dr. Charles Spon, under dates of March 10 and 22, 1648, for proof of the fact. Macaulay also says that Cardinal Mazarin was a great tea-drinker, and Chancellor Seguier, likewise.

Frankest and shrewdest among men of brains who have given to the world their inmost thoughts, old Samuel Pepys, pauses in the midst of conferences with Kings and Princes to record that "I did send for a cup of tea (a China Drink) of which I had never drank before." This in September 1660. Seven years later he writes in that wonderful Diary—"Home, and there find my wife drinking of tee, a drink which Mr. Pelling, the Potticary, tells her is good for her cold and defluxions." Then goes on to rejoice over the repulse of the Dutch in an attempt upon London.

To coffee and tea are due the establishment of that unique English institution, the London Coffee House. Inns, where quests were expected to lodge as well as eat; restaurants, in which men tarried only for a single meal; and Beer and Spirit shops, abounded in London; but the Coffee House ushered in a new era, and actually changed the daily habits of a large majority of representative London citizens. While it is asserted Mr. Jacobs established the first Coffee House in England, at Oxford, it was a native of Smyrna by the name of Pasqua Rosee who first opened a Coffee House in London, in St. Michael's Alley, Cornhill, in 1652. Hot coffee only was here dispensed, during the day and evening.

Coffee Houses soon increased in number and extended over the business districts of London. Business men quickly recognized the value of a beverage which cleared the mental vision while refreshing and stimulating both mind and body, and repaired to the Coffee House at all hours for the joint purpose of drinking coffee and transacting business with their fellows. Coffee-Houses became the Commercial Exchanges of London, and they were also the precursors of modern English Clubs. Men of affairs, Statesmen, literary celebrities, artists, naval and military officers, all repaired to the Coffee Houses to meet each other, to hear and discuss the serious topics and the light gossip of the day.

The introduction of tea gave the coffee-houses another strong hold upon their customers, and chocolate as a beverage soon followed. Among the early dispensors of these harmless hot drinks was Thomas Garway, or as written later, Garraway, whose four-story brick coffee-house on Exchange Alley, first opened in 1659, had been a rallying point for Londoners for 216 years, when it was pulled down to make room for other structures, in 1873. Garraway left a monument that has outlasted his coffee-house, in the form of a famous tea circular.

Garway's Famous Circular is so often quoted and mutilated that we print it here in full; it has no date, but it is supposed to have been printed in 1660:

AN EXACT DESCRIPTION OF THE GROWTH, QUALITY AND VIRTUES
OF THE TEA LEAF, by Thomas Garway, in Exchange Alley, near the Royal Exchange, in London, Tobacconist, and
Seller and Retailer of Tea and Coffee.

> "Tea is generally brought from China, and groweth there upon little shrubs and bushes, the branches whereof are well garnished with white flowers, that are yellow within, of the bigness and fashion of sweet-brier, but in smell unlike, bearing thin green leaves, about the brightness of Scordium, Myrtle or Sumack. This plant has been reported to grow wild only, but doth not: for they plant it in their gardens about four foot distance and it groweth about four foot high, and of the seeds they maintain and increase their stock. Of all places in China this plant groweth in greatest plenty in the province of Xemsi, latitude 36 degrees bordering up on the west of the province of Namking, near the city of Lucheu, the Island Ladrones, and Japan, and is called ' ChA.' Of this fa-

mous leaf there are divers sorts (though all one shape), some much better than others, the upper leaves excelling the others in fineness, a property almost in all plants; which leaves they gather every day, and drying them in the shade or in iron pans, over a gentle fire, till the humidity be exhausted, then put close up in leaden pots, preserve them for their drink, TEA, which is used at meals, and upon all visits and entertainments in private families, and in the palaces of grandees; and it is averred by a padre of Macao, native of Japan, that the best tea ought to be gathered but by virgins who are destined for this work, and such, 'quae non dum manstrua patiuntur; gemmae quae nascuntur in summitate arbuscula servantur Imperatori, acpraecipuis e jus dynastus: quae autem infra nasccuntur adlatera, populo conceduntur.'

The said leaf is of such known virtues, that those very nations so famous for antiquity, knowledge and wisdom, do frequently sell it among themselves for twice its weight in silver; and the high estimation of the drink made therewith hath occasioned an enquiry into the nature threrof amongst the most intelligent persons of all nations that have travelled in those parts, who, after exact trial and experience by all ways imaginable, have commended it to the use of their several countries, and for its virtues and operations, particularly as followeth, viz:

The quality is moderately hot, proper for winter and summer. The drink is declared to be most wholesome, preserving in perfect health until extreme old age. The particular virtues are these;

It maketh the body active and lusty.

It helpeth the headache, giddiness and heaviness thereof.

It removeth the obstructions of the spleen.

It is very good against the stone and gravel, cleaning the kidneys and ureters, being drank with virgin's honey, instead of sugar.

It taketh away the difficulty of breathing, opening obstructions.

It is good against tipitude, distillations, and cleareth the sight.

It removeth lassitude, and cleanseth and purifieth acrid humours, and a hot liver.

It is good against crudities, strengthening the weakness of the ventricle, or stomach, causing good appetite and digestion, and particularly for men of corpulent body, and such as are great eaters of flesh.

It vanquisheth heavy dreams, easeth the frame, and strengtheneth the memory.

It overcometh superfluous sleep, and prevents sleepiness in general; a draught of the infusion being taken, so that without trouble, whole nights may be spent in study, without hurt to the body, in that it moderately healeth and bindeth the mouth of the stomach.

It prevents and cures agues, surfets, and fevers, by infusing a fit quantity of the leaf, thereby provoking a most gentle vomit and breathing of the pores, and hath been given with wonderful success.

It (being prepaired and drank with milk and water) strengthenth the inward parts, and prevents consumption; and powerfully assuageth the pains of the bowels, or griping of the guts, and looseness.

It is good for colds, dropsys, and scurvys, if properly infused, purging the body by sweat and urine, and expelleth infection.

It driveth away all pains of the collick proceeding from wind, and purgeth safely the gall.

And that the virtues and excellences of this leaf and drink are many and great is evident and manifest by the high esteem and use of it (especially of late years) among the physicians and knowing men of France, Italy, Holland and in England it hath been sold in the leaf for six pounds (sterling) and sometimes for ten pounds the pound weight; and in respect of its former scarceness and dearness it hath been only used as a regalia in high treatments and entertainments, and presents made thereof to princes and grandees till the year 1657. The said Thomas Gaeway did purchase a quantity thereof, and first publicly sold the said tea in leaf and drink, made according to the directions of the most knowing merchants and travelers in those eastern countries; and upon knowledge and experience of the said Garway's continued care and industry in obtaining the best tea, and making drink thereof, very many noblemen, physicians and merchants, and gentlemen of quality, have ever since sent to him for the said leaf, and daily resort to his house in Exchange Alley aforesaid, to drink the tea thereof.

And that ignorance nor envy may have no ground or power to report or suggest that which is here asserted, of the virtues and excellencies of this precious leaf and drink, hath more design than truth, for the justification of himself, and the satisfaction of others, he hath here enumerated several authors, who in their learned works have expressly written and asserted the same and much more in honour of this noble leaf and drink, viz.—Bontius, Riccius, Jarricus, Almeyda. Horstius, Alvarez Semeda, Martinivus in his China Atlas, and Alexander de Rhodes in his

Voyage and Missions, in a large discourse of the ordering of this leaf, and the many virtues of the drink, printed in Paris, 1653, part x, chap.13.

And to the end that all persons of eminency and quality, gentlemen and others, who have occasion for tea in leaf, may be supplied, these are to give notice that the said Thomas hath tea to sell from sixteen to fifty shillings in the pound.

And whereas several persons using coffee have been accustomed to buy the powder thereof by the pound, or in lesser or greater quantities, which if kept for two days loseth much of its first goodness, and forasmuch as the berries after drying, may be kept, if need require, some months, therefore all persons living remote from London, and have occasion for the said powder, are advised to buy the said coffee-berries ready dried, which being in a mortar beaten, or in a mill ground to powder, as they use it, will so often be brisk, fresh, and fragrant, and in its full vigour and strength, as if new prepaired, to the great satisfaction of the drinkers thereof, as hath been experienced by many of the best sort, the said Thomas Garway hath always ready dried, to be sold at reasonable rates.

All such as will have coffee in powder, or the berries undried, or chocolata, may, by the said Thomas Garway, besupplide to their content; with such further instructions and perfect directions how to use tea, coffee, and chocolata, as is or may be needful, and so as to be efficatious and operative, according to their several virtues.

Garway's Circular embodies the redundancy of a modern legal document with the pretentious ignorance and hifaluting language of the so-called medical treatises of his day. There are many earmarks of both lawyer and doctor in this curious composition, and we can imagine the ostentatious pride with which Garway circulated the learned sense and nonsense among patrons no wiser than himself.

CHAPTER III.

HISTORICAL — Continued.

The same year that Pepys so intrepidly drank his first cup of tea in London, a tax was imposed by the English Parliament of 8 pence (16 cents) upon every gallon of tea made and sold as a beverage in England. A like tax was levied on liquid chocolate and sherbet as articles of sale. Officers visited the Coffee Houses daily to measure the quantities and secure the revenue.

In 1710 the best Bohea tea sold in London for 30 shillings or $7.00 a pound, inclusive of a government tax of $1.25 on each pound, and the consumption in England was then estimated at 140,000 lbs. per annum.

There being no authentic record or official computation of the population of Great Britain or of England previous to 1801, no comparison can be made of English tea consumption per capta with those early days.

Dr. Samuel Johnson, when taking tea with David Garrick, the tragedian, and Peg Woffington, about the year 1735, was amused at Garrick's audible complaints that the fascinating actress used too much of his costly tea at a drawing. In 1745 the British yearly consumption of tea was but 730,000 lbs. The Scotch Judge, Duncan Forbes, in his published letters of that period, wrote that the use of tea had become so excessive, that . . .

"the meanest families, even of laboring people, particularly in boroughs, make their morning's meal of it, and thereby disuse the ale which heretofore was their accustomed drink; and the same drug supplies all the laboring women with their afternoon's entertainment, to the exclusion of the twopenny," (i.e., dram of beer or spirits).

So that we may trace our ultra-fashionable 5 o'clock tea of 1900 back to its plebian origin among plain working people, to the working woman, to the washerwoman of 150 years ago. Let the revived custom not lose caste by this admission, but rather gain in wholesome popular estimation by evidence of a common tie between the humblest and the most fortunate of mankind.

A president of an English Court of Sessions also complained that tea was driving out beer, and indirectly injuring the farmer, in whose cottage, he omitted to say, the tea canister had begun to occupy a place of honor, despite the lessened demand for his malt.

In 1745, the British tea tax was reduced to 1 shilling (25 cents) per pound, together with 25 per cent of the gross price. The selling price immediately dropped, and British consumption in 1846 rose to 2,358,589 lbs. The use of tea has often been checked by excessive duties or excise tax. From 1784 to 1787 British consumption rose from five million pounds to seventeen millions of pounds, consequent upon a reduction of duties. Twenty years after, under the imposition of exorbitant duties, British consumption was only nineteen and one quarter millions of pounds.

It was in those early years of the nineteenth century that tea firmly and permanently established itself in the humbler households of England. Its economical prominence elicited from William Cobbett, the economist and pugnacious editor, a declaration that from eleven to twelve pounds of tea constituted the average annual indulgence of a cottager's family, at a cost of eight shilling for black and 12 shillings for green tea ($2 to $3) per pound, which was doubtless an over-estimate. And we must bear in mind that tea in those days was sold by the ounce, measured into the teapot by the grain, and was steeped until every vestige of flavor, savory or bitter, had been extracted from the precious leaves.

Although in 1807 the governing powers of Great Britain forced excise duties on teas up to ninety per cent. of their cost, tea had been proved to be so beneficial and essential to happiness by British workers that Charles Dickens, in reviewing the situation, presents it as follows:—"And yet the washerwomen looked to her afternoon 'dish of tea' as something that might make her comfortable after her twelve hours of labor, and balancing her saucer on a tripod of three fingers, breathed a joy beyond utterance as she cooled the draught. The factory workman then looked forward to the singing of the kettle, as some compensation for the din of the spindle. Tea had found its way even to the hearth of the agricultural laborer, and he would have his ounce of tea as well as the best of his neighbors." But the heavy taxed worker was often forced to choose between a tea adulterated with English plants of other kinds, or the contraband but genuine commodity offered by enterprising smugglers, who were the despair of the Crown officers of the revenue, and the recognized friends of the over-taxed poor.

It must not be inferred that tea as a beverage became naturalized in England without meeting with the unreasoning opposition that usually greets the advent of a stranger. The press and pamphlets of the day contained frequent attacks upon tea, and the violence of denunciation usually bore a fair proportion to the ignorance of the writer; ignorance of physiology, ignorance of medicine, ignorance of the pamphlets itself. The unfavorable opinions and portentous predictions of some of the physicians of the period are among the curiosities of medical records. Tea, like all other things, may be abused, and a good friend be converted into an enemy. But cold water has killed many persons, and plain bread sometimes proves indigestible.

The plant whose leaves yield the tea of commerce is variously termed Camellia Theifera; Thea Sinensis; or Chinensis; Thea Assamica; Thea Bohea and Thea Viridis, according to its origin, variety of the writer's fancy. While the real character of the East Indian or Assam tea plant has been recognized by botanical science less than seventy years, and the Chinese tea plant has probably been utilized for fifteen hundred years, it will be more convenient to begin our remarks with the later discovery.

Writers at the present time continue to describe the tea plant as a "shrub" of about six feet in height. The indigenous tea plant of India, which is believed to be the parent stock of Chinese tea plants, is a tree, growing to a height of 20 to 35 feet with a trunk 8 to 10 inches in diameter, and bearing leaves of a lively green, 8 to 9 inches in length and 4 inches in breadth. The leaves are much more delicate in texture than those of Chinese plants, which hardly reach 4 inches in length, and the former contain a larger percentage of the invaluable alkaloid, Theine. Dr. Chas. U. Sheppard, in a historical sketch of Tea Culture in South Carolina, tells us that a tea tree which was planted planted by Michaux, about 15 miles from Charleston, and about the year 1800, had attained a height of say 15 feet when he saw it a few years ago.

The native Indian tree is, however, not now utilized upon a commercial scale for tea purposes. The reason for neglecting the native plant we do not find definitely stated, but infer from several sources of information that it is owing to the extreme delicacy of constitution of the Assam plant, its demands for excessive moisture and high temperature, and its preference for partial shade, evidenced by its growing in the jungle and under other trees. Possibly a difficulty in restraining its luxuriant habit of growth is also involved. However this may be, the commercial tea of Ceylon and India is a product of a cultivated cross between the tender native Indian and the hardier Chinese tea plants, in which the Assam strain bears the proportion of one half to two thirds. A more robust plant under cultivation is the result, and one which preserves the best qualities of both varieties. This cross is usually termed a hybrid.

It seems probable that the removal of the tropical Indian plant to China, more than a thousand years ago, with its much colder and dryer climate and its poorer soil—for the best soil of China has been set apart for rice and other indispensable foods— together with continual removal of its leaves, have in time evolved a tea plant so different from its parent stock, that scientists failed for many years to recognize the Indian original. Several times in the early years of this century zealous travellers and residents of India sent to England specimens of the native Indian tea plant for scientific examination. But conservative government officials had already established a botanical or technical standard for the tea plant to which every aspirant for relationship must conform; no one of them seems to have thought of the simple test of the teapot. Finally some rash investigator, not having the fear of scientific anathema before his eyes, crudely cured a few leaves, and actually put them in hot water. Tea merchants immediately recognized the plant and the magic circle of the Circumlocution Office was smashed into bits.

Meanwhile, Chinese tea plants and Chinese experts and laborers had been imported into India and tea gardens were well under way before the native tea plant had been recognized. But in the ultra-tropical climate of India, Chinese tea plants languished, and success was finally obtained only by abandoning the stunted Chinese varieties, and getting back nearer to the indigenous Thea Assimica; and by the introduction of modern agricultural methods under British management, and even by the use of machinery for rolling tea and for firing tea by currents of hot air. Indian laborers now supersede the Chinese workmen, who were not found sufficiently pliable in adapting themselves to European ideas.

To preserve the historical record of tea so far as possible, we will state that while the indigenous Indian tea plant had been recognized somewhere about the year 1820, the first serious and sustained attempts to grow tea in India were made by Englishmen, about 1834, using Chinese tea plants and Chinese workmen for the purpose. English authorities differ upon the exact dates. The first shipment of English grown tea from India to London was made in 1838; it amounted to but 60 chests, which brought at auction in London $2.25 a pound. The second shipment in 1839 of ninety five chests brought $2.00 a pound. In 1899 the Indian tea crop amounted to about 175,000,000 lbs., and the size of Indian tea gardens varied from 100 acres or less up to 4,000 acres. In 1897 the total acreage of tea plantations in India was stated by Mr. Crole at 509,500 acres, equal to nearly 800 square miles.

Ceylon began to grow tea on a commercial scale as late as 1875, after her coffee plantations had been ruined by disease. That year her total acreage was about 1,000 acres, In 1883 Ceylon exported a million and a half pounds of tea. In 1897 she had 400,000 acres of growing tea, equal to 625 square miles; and the estimate of Mr. William MacKensie, Tea Commissioner for the Ceylon Government, of her production for 1900, is 135,000,000 lbs.

The aggregate exports of tea by India and Ceylon is about 310,000,000 lbs., a complete reversal of conditions of tea trade within twenty years, and due entirely to British enterprise and the fine quality of British grown teas.

A liberal estimate for the total exports of Chinese and Japanese teas for 1899 would be 340,000,000 lbs.; so that it is fair to say that the world's consumption of tea, outside of China and Japan, is now equally divided between teas of the latter two countries and those of English growth.

CHAPTER IV.

Characteristics Of The Tea Plant.

Chinese tea plants are usually divided into two classes, and distinguished a Thea Bohea and Thea Viridis, the former being most suitable for black teas, and the latter for green teas; and black and green teas have been indiscriminately made from the leaves of either.

A tea shrub of Chinese origin now before us, growing among a host of common American plants, displays no special characteristics which would attract attention to itself. It resembles an orange plant. Its developed leaves are smooth on the surface, leathery in texture, dark green in color, with edges finely serrated from point almost to stalk. They are without odor, and when chewed in the mouth, have a mild and not unpleasant astringency, but no other perceptible flavor. A leaf of any familiar domestic plant, such as the lilac, the plantain, or the apple, has a stronger individuality to the sense of taste, than this green leaf of the tea plant.

How was the hidden mystery of its incalculable value to mankind revealed? What premonition guided the Chinese discoverer to the preparatory treatment and delicately graduated firing process which develops tea's precious flavors? And does not this unsolved question suggest the possible existence of other plants, growing, perhaps, at our very doorsteps, possessing rare and unrecognized virtues?

In form, tea leaves have been compared by writers to leaves of the privet, the plum, the ash, the willow, but close observers know that not only do leaves of the species just mentioned represent different types, but that important variations in form occur in leaves of the same species, and in leaves growing on a single tree or plant. The tea plant is subject to the same vagaries, and any description by comparison will be misleading. The reader must be content with the typical forms of tea leaves shown in our engravings on the following page, for which we are indebted to the kindness of Mr. Joseph M. Walsh, importer of teas, at Philadelphia.

All varieties of the tea plant bear a pure white flower, averaging, say 1 1/4 inches in diameter, and resembling very closely our single white wild rose blossom.

Its bunch of bright yellow stamens is so bushy and showy in some varieties that careless travelers have been led to report the flower as yellow in color, which is never the case.

In some Chinese plants, and in those of India, tea blossoms are very fragrant, and they have been used for scenting tea leaves in India, if not in China, as other flowers are used by the Chinese. In India a perfume has been distilled from tea blossoms; and a valuable oil is expressed from the very oily seeds. The long tap root of the tea plant renders it difficult to transplant.

In China, tea is commonly cultivated in small patches or fields, large tea fields being the exception. The nature of Chinese inheritance laws and customs which tend to continual subdivision of land, may be one of the causes of this state of affairs. The least area of spare ground is frequently utilized by the small farmer or the cottager for the cultivation of a dozen or more tea shrubs, from which they procure tea for their own use, or realize a small sum by sales of the green leaves to tea traders. Many a rocky hillside or mountain slope, otherwise waste ground, is terraced so as to detain the rains and meagre soil within its inwardly inclined banks and trenches, and made to yield a valuable crop of tea. Indeed, some of the finest flavored Chinese tea, of fabulous value where they are produced, are grown in seemingly inaccessible retreats among precipitous mountains.

The plate on the following page is a reproduction of a Chinese drawing brought from China by Robert Fortune, the Scotch botanist and traveler, and first published in Mr. Fortune's Two Visits to the Tea Countries of China, London, 1853, now out of print. The picture represents with Chinese fidelity a scene on the River of Nine Windings, in the Bohea Hills, and in the heart of a black tea district. Mr. Fortune spent several days at the scene of the illustration, and writes of the country as follows:

"Our road was a very rough one. It was merely a foot path, and sometimes narrow steps cut out of the rock. When we had gone about two miles we came to a solitary temple on the banks of a small river which here winds amongst the hills. This stream is called by the Chinese, the river of the Nine Windings, from the circuitous turnings which it takes amongst the hills of Woo-e- shan. Here the finest Souchongs and Pekoes are produced, but I believe that they rarely find their way to Europe, or only in small quantities. The temple we had now reached was small and insignificent building. It seemed a sort of half way resting place for people on the road from Tsin-Tsun to the hills, and when we arrived, several travelers and coolies were sitting in the porch, drinking tea. The temple belonged to the Taouists, and was inhabited by an old priest and his wife. . . . The old priest received us with great politeness, and according to custom gave me a piece of tobacco and set a cup of tea before me. Sing- Hoo now asked whether he had a spare room in his house, and whether he would allow us to remain with him for a day or two. He seemed very glad of the chance to make a little money, and led us up stairs to a room. The house and temple, like some which I already described, were built against a perpendicular rock which formed an excellent and substantial back wall to the building. The top of the rock overhung the little building, and the water from it continually dripping on the roof of the house gave the impression that it was raining.

"The stream of the Nine Windings flowed past the front of the temple. Numerous boats were plying up and down, many of which, I was told, contained parties of pleasure who had come to see the strange scenery amongst these hills. The river was very rapid, and these boats seemed to fly when going with the current, and were soon lost to view. On all sides the strangest rocks and hills were observed, having generally a temple and a tea manufactory near their summit. Sometimes they seemed so steep the the buildings could only be approached by a ladder; but generally the road was cut of the rock in steps, and by this means the top was reached. . . .

Some curious marks were observed on the sides of some of these perpendicular rocks. At a distance they seemed as if they were the impress of some gigantic hands. I did not get very near these marks, but I believe that many of them have been formed by the water oozing out and trickling down the surface; they did not seem to be artificial; but a strange appearance is given to rocks by artificial means. Emperors and other great and rich men have had stones with large letters carved upon them let into or built in the face of the rocks. At a distance these have a most curious appearance. . . .

I now bid adieu to the famous Woo-e-shan, certainly the most wonderful collection of hills I ever behold."

He says further that some geologist who will visit the scene, may "give us some idea how these strange hills were formed, and at what period of the world's existence they assumed the strange shapes which are now presented to the traveller's wondering gaze."

CHAPTER V.

Tea Picking And Yield.

Chinese tea grown among the mountains and hillsides was in Mr. Fortune's time distinguished as "Hill tea," while both large and diminutive plantations on the lowlands or the plains were all called "tea gardens," a term which is now applied by the English to the extensive plantations of Ceylon and India.

Some of the largest tea plantations in China turned out, say, 500 chests, or 30,000 pounds, of tea per annum, at the same period.

In both China and the East Indies a common custom prevails of planting tea bushes about four feet apart, each way, and they are pruned down to a height varying from three to six feet, to bring the topmost leaves within reach of the picker. In both named countries, a first crop of tea leaves may be gathered from the plant at three years from the seed, but a full crop is not expected until the plant is about six years old. "A Chinese plantation of tea, seen from a distance," says Mr. Fortune, "looks like a little shrubbery of evergreens." And when journeying in the Bohea black tea country, he remarks—"As we threaded our way amongst the hills I observed tea gathers busily employed on all the hill sides where the plantations were. They seemed a contended and happy race; the joke and merry laugh were going around; and some of them were singing as gaily as the birds in the old trees about the temples." There is an old Chinese ballad of some 30 stanzas, which pictures the reflections of a Chinese maiden who is employed in picking tea in early spring, from we select a few verses, literally translated.

"Our household dwells amidst ten thousand hills,
 Where the tea, north and south of the village, abundantly grows;
 From Chinshe to Kuhyu, unceasingly hurried,
 Every morning I must early rise to do my task of tea.

"By earliest dawn, I at my toilet, only half dress my hair,
And seizing my basket, pass the door, while yet the mist is thick;
The little maids and graver dames hand in hand winding along,
Ask me, 'which steep of Sunglo do you climb to-day?'

"My splint-basket slung on my arm, my hair adorned with flowers,
I go to the side of the Sunglo hills, and pick the mountain tea.
Amid the pathway going, we sisters one another rally, And
laughing, I point to younder village—'there's our house!'

"This pool has limpid water, and there deep the lotus grows;
Its little leaves are round as coins, and only yet half blown;
Going to the jutting verge, near a clear and shallow spot,
I try my present looks, mark how of late my face appears.

"The rain is passed, the utmost leaflets show their greenish veins;
Pull down a branch, and the fragrant scent is diffused around.
Both high and low, the yellow golden threads are now quite culled;
And my clothes and frock are dyed with odors through and through.

"The sweet and fragrant perfumes like that from the Aglaia;
In goodness and appearance my tea'll be the best in Wuyen,
When all are picked, the new buds by next term will again burst forth,
And this morning, the last third gathered is quite done.

"Each picking is with toilsome labor, but yet I shun it not,
My maiden curls are all askew, my pearly fingers all be numbed;
But I only wish our tea to be of a superfine kind,
To have it equal their 'dragon's pellet,' and his 'sparrow's tongue.'

"For a whole month, where can I catch a single leisure day?
For at earliest dawn I go to pick, and not till dusk return;
Then the deep midnight sees me still before the firing pan—
Will not labor like this my pearly complexion deface?

"But if my face is thin, my mind is firmly fixed
So to fire my golden buds that they shall excel all beside,
But how know I, who'll put them in jewelled cup?
Whose taper fingers will leisurely give them to the maid to draw?"

Men, women and children are in China employed for picking tea, and three crops are gathered in favorable seasons, with occasionally a fourth picking. Under the stimulus of East Indian heat and moisture, the "flushes," or new growth of shoots, buds and leaves, are renewed as often as once in a week or ten days; so that during a season of nine months, from a dozen, to a maximum of thirty pickings are made. The same conditions apply to the tea plantations of Java. After ten or twelve years the bushes decline in vigor from the strain of constant loss of young growth, and are replaced by new plants. Thirty pounds of green leaves are an average day's work for women and children.

The yield of green leaves or of cured dry tea from a single bush is necessarily variable with its age, size and condition. In China, the proportion of manufactured tea to the green leaves is one to three, or one to three and one-third, while in the East Indies and Java the allowance is one to four.

Statistics gathered from India tea planters give us the following figures, for different districts and years:

YIELD OF DRY TEA PER ACRE, PER ANNUM.
Pounds………….. 370 333 330 246 562

YIELD OF DRY TEA PER BUSH, PER ANNUM.
Ounces………….. 1.18 1.46 1.44 1.08 2.50

Mr. Owen A. Gill, of Messrs, Martin Gillett & Co., Baltimore, in 1891, estimated the yield of Indian tea plantations at 400 pounds per acre per annum, costing at that time in India, ready for shipment, say, ten cents a pound; to which must be added, freight, selling charges, etc., of at least four cents a pound.

Half century ago, Mr. Fortune estimated that in China the small grower realized for a common Congo tea, about four cents a pound, but that boxing, transportation to the coast, export duty, etc., brought the cost in Canton to about ten cents a pound. Fine teas then paid the grower, say, eight cents a pound, but the English merchants in Shanghai paid thirty cents for the same teas.

Dr. Charles U. Shepard of the Pinehurst tea plantation at Summerville, S.C., recently stated that Chinese bushes are said to produce 2 ounces of dried tea per bush; those of Japan, 1 ounce per bush or less; those of India and Ceylon averaging 3 to 4 ounces, and on high ground, 2 to 3 ounces; while Dr. Shepard has gathered from his own plantation, from acclimatized Assam crosses, 3 ounces per bush, and from Chinese plants, 4 to 5 ounces. His Japan plants yielded but 1/2 ounce of tea.

Picking tea on the level lands of India and Ceylon is very light work, and women and children are almost exclusively employed. Mr. David Crole, writing in the serious and practical vein of a scientific expert, is moved to a poetic sense of the scene when he speaks of the return of Indian tea pickers from their work at evening:—

"A long line of women with their gay clothes of various hues, lit up by the expiring gleams of the setting sum, winding their way along the garden paths, like some monster snake, with scales of many colors; their gait perfect, undulating, and undisturbed by the baskets poised gracefully on their heads; singing some quaint refrain in the usual minor key, or making the air gay with their chatter and laughter; which, if far distant, strikes the ear pleasantly as a faint and indistinct hum."

The tea plant undoubtedly reaches its highest perfection as a member of the vegetable kingdom, in India and Ceylon, in a climate of extreme heat and extreme rainfall and moisture, and in a very rich soil; and the remark is often heard from Indian planters that "tea and malarious fevers flourish together." Experience has shown however that the tea plant possesses a wonderful power of accomodation to adverse conditions. In China and in the United states, it has been taught to put up with a comparatively sterile soil, dry mountain air, at heights in China reaching 6,000 feet above sea level, and occasional temperatures as low as 12 to 10 degrees Fahr., in the midst of recurrent ice and snow.

The story of tea in Japan alone calls for more space than this entire book could furnish, and there is an ample field for a treatise upon the cultivation, preparation, and social importance of tea in that strikingly interesting land. Nearly one half of the tea consumed in the United states comes from Japan, our imports of Japan tea being about 44,000 pounds during last year. Although tea has been grown in that country for more than siz centuries only about forty years.

Tea in Japan is largely grown upon hill-slopes and in small plantations or gardens, the latter term being peculiarly appropriate to their neat, symmetrical and picturesque appearance. The character of the soil is noticeably connected with the quality of the tea. From the putting forth of new leaves in the Spring-time until the advent of its white fragrant blossoms in the Autumn, the tea plant is an object of admiration and affection with the susceptible, nature-loving Japanese.

We are indebted to an English gentleman and tea merchant who has resided in Japan for 30 years, for many interesting facts connected with our subject.

He tells us that while the principal crop of teas for export is produced on plantations of comparatively recent establishment, there are tea gardens in the interior of Japan which have been cultivated for 500 years; and that tea is still gathered from bushes which spring from roots which were planted 100 to 300 years ago. These ancient plants yield a tea in limited quantities which is elaborately and expensively prepared for the nobility and wealthy Japanese, and commands prices running up as high as ten dollars a pound. Some of the choice tea which comes to this country is picked from plantations which have been in existence for 300 years, and is sold under the names of "challenge," "Violet," and "Japonica" teas.

These facts are in striking contrast with the limited life of Chinese tea plants, as stated by Mr. Fortune.

Japan teas do not fall into either of the three classes into which Chinese and Indian teas have been divided. They have been styled green teas by the trade, but that appelation grew out of their customary color, and their mild odor and taste; while Japan Black teas are now produced from the same leaf. Japan teas are favorites with many persons who do not relish the herby taste of other Black teas, and with whom Chinese Green teas disagree.

CHAPTER VI.

Tea Manufacture.

The tedious, long-drawn-out details of tea manufacture, of the repeated, meaningless, tossing back and forth and Chinese juggling with the abused tea leaves, are but too familiar to students of the subject: and too disappointing also, when we are moved to ask— Why all this manipulation? What is the nature of the chemical changes which take place?

So far as we can ascertain by diligent inquiry and reading, no competent authority has answered these questions satisfactorily. We have been deluged with generalities and opinions which contradict themselves, but when we search for a categorical answer to a simple question, experts hide under a shower of meaningless phrases. We, alas, are not an expert, nor a chemist, but just a simple enquirer in search of knowledge expressed in plain English. Therefore be patient dear reader with our endeavors to represent or interpret existing conditions of expert knowledge of tea manufacture at this time. Peradventure a feeble ray of light may illuminate the darkness of the subject. Corrections and additions will be welcomed in our future editions and credit given to their authors.

Teas may conveniently be divided into the three classes which have so long been recognized by the American tea trade, namely:

Green teas, the first remove from the green leaf.
Oolongs, delicate Black teas, having properties further developed than those of Green teas.
Souchongs, and Congous, both of which have been called "English Breakfast" teas by Americans, because the former teas were the customary breakfast beverages of the English people before the advent of Indian teas.
In these latter teas, fermentation and firing are prolonged beyond the treatment of Oolongs. The smoky flavor sometimes apparent is owing to careless and extreme firing.

In making Green tea, the object seems to be to expel the watery juices of the leaf and to cure or dry it with the least delay. Hence, the leaves are not exposed to the sun, but are first dried in the air for a short time. They are next exposed to artificial heat, which renders them flaccid and pliable, and prepares them for the third operation of rolling, which twists the yielding leaf as seen in manufactured tea, rolls it up into balls, and squeezes out a considerable portion of its watery juices. It is a singular fact that in the Chinese methods, they endeavor to get rid of the exuding juices, while in the Indian treatment, according to Mr. crole, the manufacturing expert, effort is made to preserve the sappy juice, and it is continually taken up again by the balls of leaves. The balls are now broken apart, and the scattered leaves are submitted to the final drying process by fire, which finishes Green tea. In this case, it is plainly the heating treatment which develops the faint flavor and odor of Green tea, for no fermentation is allowed to begin, unless indeed brief and unobserved action takes place within the compressed balls.

In making an Oolong Black tea, which occupies an intermediate position between Green tea and Black Souchongs and Congous, the leaves are first exposed to the action of the air for a considerable time, and in many cases, to the sun also. An incipient fermentation may take place, although this is denied by some. There is certainly a chemical change beyond the brief preliminary drying of Green tea. During this period the leaves (in China) are stirred and tossed by the hands. The effect, if not the object, is to expose greater surfaces to the air, and to increase oxidation. It is during this operation that the leaves first begin to manifest characteristics of manufactured tea, in the way of a fragrant tea odor which the green leaf did not possess. The development of sweet odors in new hay, quite different from those of green grass, and also the artificial development of flavor in tobacco leaves, may be recalled in this connection. This prolonged exposure to the air is termed "withering," and the leaves become soft and flaccid, as they do in the first artificial heating for Green tea. In withering, the leaves lose about one quarter of their weight in moisture. The leaves must not be bruised before the termination of this treatment, or injurious chemical changes will begin.

The second operation with Black tea is the same rolling into balls, twisting and squeezing, as in Green tea. Mr. Crole says that the sap of the leaf thus liberated from its cells "is spread all over the surface of the rolled leaf, where it is in a very favorable position for the oxygen of the atmosphere to act upon it during the next stage of manufacture, namely, fermentation." Fermentation, he regards as an oxidation process mainly.

For the "fermentation" stage, if that controverted term correctly designates the process, the rolls are either left undisturbed to heat, or, as in Indian methods, the rolls are broken up, and the leaves distributed in drawers, with free access of air. In either case, a spontaneous heating follows, and chemical action is indicated by a change of color which reddens and darkens the leaf, and by the evolution of further pleasant "tea" odors. Some of the tannin is said to be converted into glucose.

Care must be taken, Mr. Crole says, to arrest fermentation at the proper stage by the first "firing," and this firing expels about half of the remaining moisture of the withered leaves, and probably develops an additional portion of those volatile oils which give fragrance and taste to manufactured tea; and which Mr. Crole designates by the name of "theol." Too high or too long continued firing drives off these oils with the watery juices. They are also wasted by exposure of manufactured tea to the atmosphere. Firing is sometimes divided into two or three stages.

In the above summary we have described all essential treatment of tea leaves necessary to produce manufactured tea.

To procure the extreme type of Black teas, a Souchong or Congou, the fermentation or oxidation, and the "cooking" process, is simply carried further, and with higher roasting, some of the volatile oils and delicate flavors are expelled, or are changed into other flavors. Judging by diminished effects upon tea drinkers, some of the volatile theine is also lost.

Both in China and Japan it is the custom to give large portions of the tea crop which are intended for export to foreign countries, only a preliminary drying or curing sufficient to preserve them temporarily. When they arrive at the shipping ports they are subjected to additional firing and thorough drying.

CHAPTER VII.

Chemistry and Physiological Aspects of Tea.

If the reader desires an example of imperfect and arrested knowledge in some of the common affairs of life, let him collate the statements of scientific experts concerning the physiological effects upon mankind, of tea. He will then admit that "in a multitude of counsellors there is confusion."

Without pretending to more than the rudiments of chemical or physiological science, we shall attempt to examine the nature of tea, and its effects upon the human system; taking as a basis for our remarks Professor Jas. F. Johnston's Chemistry of Common Life, from which work more recent writers draw most of their inspiration.

Chemists find in manufacturing tea leaves three principal constituents to which all the physiological effects of tea are attributed. These are, (1) Theine, (2) Essential or Volatile Oils, (3) Tannin.

Theine is present in the green leaf of tea, and is apparently unchanged in the manufactured leaf and in the infusion or beverage. We regard it as the one essential and the most valuable element of all teas, physiologically considered. Strangely enough theine is the one important constituent which is entirely neglected by the tea-tester and the trader. In testing and grading teas for purchase and sale, their appearance, odor and taste, their color and body when "drawn," determine their pecuniary value, without relation to their percentage of theine, or its effects upon the tester.

Theine has been found in nature in but a few plants, as in tea, in coffee, (then termed caffein), in Mat'e (Paraguay or Brazilian tea), and in the Kola nut of Africa. A very similar principle, having analogous properties, but containing more nitrogen, exists in cocoa, (theobroma).

Theine, when isolated by heat from the tea leaf or infusions, condenses in minute white needles or crystals, having no odor and but a faintly bitter taste. In manufactured tea leaves, theine constitutes from one to five percent. of their weight. According to Professor Johnston, three or four grains per day of this substance may be taken without injury by most persons; or such quantity as would be contained in half and ounce of Chinese black tea. Indian (Assam) tea and Ceylon tea, being stronger in theine, would suffice in lesser quantity.

Theine is soluble in about 100 parts of hat water. It vaporizes at 185 degrees C. or 365 degrees Fahr., hence it is not driven off by continued boiling of tea infusion.

W. Dittmar found by experiment that prolonged steeping of tea leaves up to ten minutes increased the proportion of theine in the infusion. His results are as follows:

STEEPED 5 MINUTES.

Average of 8 samples Chinese tea:

Theine, per cent infusion—2.58 Tannin—3.06

Average of 6 samples Ceylon tea:

Theine—3.15 Tannin—5.87

Average 12 samples of Indian tea:

Theine—3.63 Tannin—6.77

STEEPED 10 MINUTES.

Theine, per cent infusion—2.79—Increase about 10 per cent
Tannin—3.78—Increase about 25 per cent

Average of 6 samples Ceylon tea:

Theine—3.29—Increase about 5 per cent Tannin—7.30—Increase about 25 per cent

Average 12 samples of Indian tea:

Theine—3.73—Increase about 3 per cent Tannin—8.09—Increase about 20 per cent

W. M. Green reported that in prolonging the steeping of tea from 10 to 20 minutes, he observed the formation of a tannate of theine, which diminished the proportion of 1.30 per cent. of theine at 10 minutes to 1.16 per cent. after 20 minutes steeping, a loss of about 10 percent., unless the latter salt so formed is proved to yield up its theine constituent in the human stomach.

While theine is credited as the source of the most powerful and useful properties of tea, and without which no plant would be recognized as tea, yet some of the stimulating or exhilarating influences of this plant are attributed to the volatile oils which contribute so largely to the flavors and odors which characterize tea.

These Essential or Volatile Oils of manufactured tea are said to reside in the minute cells of the green leaf, but they are greatly changed by manipulation, for they are not manifest to the sense of taste or smell when expressed from the green leaf by bruising, nor does the green leaf yield their aromatic flavors to an infusion. Professor Johnston says that these precious oils are artificially developed by manufacture. David Crole declares that they are developed "to a certain extent during withering, and also during the first stage of firing," which last process, if carelessly conducted, "oxidises it (the oil) into resin."

Green tea, they first remove from the green leaf, imparts very little flavor or scent to its infusion. In some Oolong Black teas, and in some Ceylon Black teas, these oils are highly developed and are very fragrant. In the black Souchongs and Congous they have again been altered by treatment, but are no less perceptible, and to many, are quite as agreeable. Although constituting only one-half to one per cent. by weight of the dried leaf, these oils are all-important to the trademan and to the consumer.

These volatile oils are strongest in new teas, and are gradually wasted by exposure to the atmosphere. Robert Fortune and other travelers in China have stated that the Chinese will not use new teas, but allow them to pass through a sort of "ripening" process. Mr. Crole, speaking probably of the Indian teas with which he was so familiar as a planter and chemist, says that "tea should always be kept for a year before being drank. If the infusion of freshly manufactured tea is drank, it causes violent diarrhea; therefore it should be kept a year before it is consumed, in order to let it mellow."

There is no doubt that the more impervious the package containing tea is to the air, the more perfectly the finer qualities of the tea are preserved. If there is a necessity for ripening or mellowing by time, air should be rigidly excluded during that period.

As to the keeping qualities of fine teas, in tight packages, we know that they are not spoiled or injured by two years storage in this climate.

Tannin is the third important element of the tea leaf, and it varies greatly in percentage in different teas, and increases with the age of the growing leaf. It is the cause of the rasping, puckering, astringent effect upon the tongue and interior of the mouth.

Tannin in tea has been a great bugbear with the ill-informed, bit it is not nearly so deleterious as some careless or unscrupulous writers would have us believe. In the first place there is a very insignificant quantity of tannin in properly drawn teas, say in those drawn for not longer than five or eight minutes. The tannin present in a fine Black tea, steeped at a moderate temperature for fifteen or twenty minutes will not harm a delicate stomach. We take quite as much tannin in some fruits, and make no fuss about it. Secondly, if a strong solution of tannin is taken into the stomach and there comes in contact with albuminous or gelatinous foods, it will expend its coagulating power upon such substances. If there are no such substances present, it is the expressed opinion of Mr. Crole (in a discussion upon the chemistry of tea) that the tannin is converted into glucose and other harmless products by the digestive processes. The wild declarations that tea tannin "tans" the coating of the stomach into a leathery condition is without foundation. Even where too prolonged steeping has greatly increased the usual proportion of tannin in tea infusion, milk, when added, neutralizes the coagulating power of the tannin entirely or to such degree as to render it harmless.

Professor Johnston thinks it quite probable that tannin takes some part in the exhilarating effect of tea, and in that of the betel-nut of the East. While the astringent influence of strong tannin upon the bowels is regarded as unfavorable, hot tea infusion has with many persons a contrary effect, stimulating the peristaltic movements and antagonizing constipation.

If tannin is injurious, it should be observed that its proportion in the leaf of green teas is very much larger than in Black teas. An analysis by Mulder gave as the percentage of tannin in a Black tea, 12.85 per cent., and in a green tea as 17.80 per cent. But another analysis made by Y. Kazai, of the Imperial College of Agriculture of Japan, made the per centage of tannin (gallo- tannic acid) in a Green tea 10.64, and in a Black tea from the same leaf 4.89. In the green leaf from which these teas were derived he found 12.91 per cent. of tannin. This analysis indicates also that a portion of the tannin disappears in manufacturing Green tea, but a still larger, proportion is lost or changed in the manufacture of Black tea.

Tannic acid taken into the human stomach in large quantity produces, according to the U.S. Dispensatory, "only a mild gastrointestinal irritation."

Passing over the phosphoric acid, the gluten, and other interesting constituents of the tea leaf, we proceed to the observed effects of tea upon the human system.

Professor Johnston (before quoted) says that tea "exhilarates without sensibly intoxicating. It excites the brain to increased activity and produces wakefulness; hence its usefulness to hard students, to those who have vigils to keep, and to persons who labor much with the head. It soothes, on the contrary, and stills the vascular system, (arteries, veins, capillaries, etc.), and hence its use in inflammatory diseases, and as a cure for headaches. Green tea, when strong, acts very powerfully on some constitutions, producing nervous tremblings and other distressing symptoms, acting as a narcotic, and in inferior animals even producing paralysis. Its exciting effect upon the nerves makes it useful in counteracting the effects of fermented liquors, and the stupor sometimes induced by fever." And again, tea "lessens waste," and diminishes the quantity of food required; "saves food; stands to a certain extent in the place of food, while at the same time it soothes the body and enlivens the mind."

Professor A. H. Church, of Oxon, England, in one of his often quoted books on Food, says that "the infusion of tea has little nutritive value, but it increases respiratory action, and excites the brain to greater activity."

J.C. Hutchinson, M.D., (late President Medical Society of State of New York), remarks that caffein, which he regards as identical with theine, "is a gentle stimulant, without any injurious reaction. It produces a restful feeling after exhausting efforts of mind or body; it tranquilizes but does not disqualify for labor, and therefore it is highly esteemed by persons of literary pursuits. The excessive use of either tea or coffee will cause wakefulness."

Dr. Kane, the Artic Explorer, speaking of the diet of his men while sojourning in the Artic ice fields, said that his men preferred coffee in the mornings, but at night, "tea soothed them after a hard day's labor, and better enabled them to sleep."

Dr. Edward Smith, an English Physiologist, in an address before the Royal Medical and Chirurgical Society, remarked that "tea increased waste in the body, excited every function, and was well fitted to cases where there was a superfluity of material in the system;—but is injurious to the under-fed, or where there is greater waste than supply." Dr. Smith recommended tea as a preventive of heat-appoplexy, and in cases of suspended animation, as from partial drowning.

We have selected these expressions of opinion from among a large number of diverse character, for the purpose of illustrating the uncertainty of knowledge concerning tea. To recapitulate:—

Professor Johnston finds that tea exhilarates; excites to activity, produces wakefulness; yet it sooths, and it tranquilizes the vascular system; it lessens waste and saves food.

Dr. Smith found tea to increase waste, and to be injurious where food is deficient; says tea excites every function,—which must include the vascular system.

Dr. Hutchinson and Dr. Kane agree in the main.

What is the meaning of such radical differences of view? We think they arise from three causes: First, tea affects different persons very differently; secondly, the subject has not received that careful study which it merits, and thirdly, there is a careless confounding of at least three classes of effects, and a confusion of terms in describing them.

We feel an unaffected diffidence in criticising and endeavoring to improve upon the expressions of scientific men of honest purpose, but we may be pardoned for pointing the way to a more careful analysis of the merits and deficiencies of an article of diet used by so many millions of people.

We find among the ordinary effects of tea-drinking:

Exhilaration: — an elevation of feeling, a lightness of mood or spirits; a cheerfulness or even joy, which is compatible with rest. This effect may be entirely independent of pure stimulus, or of any disposition to mental or physical activity.

Stimulation: — a quickening or rousing to action of any faculty, but as usually employed, an urging to action of bodily or mental powers.

Sustaining: — enabling one to continue the expenditure of energy with less sense of fatigue, at the time, or afterwards.

Refreshing: — relieving or reviving after exertion of any kind; reanimating, invigorating; contributing to rest after fatigue.

Exciting: — in the sense of stimulation of brain and nervous system to higher tension, but not necessarily attended by disposition to labor or useful activity.

Now some tea-drinkers find in the beverage exhilaration only, a lightness of mood, but they are disposed to rest and to revery, to simply a passive meditation, or an indulgence of the imagination.

Others are stimulated to mental or to physical activity, and are sustained during such action. Afterwards they are refreshed when fatigued, by the same beverage.

Others again are nervously excited and cannot rest or sleep; but are too "nervous," as they express it, to set about any formal task, especially of a mental character.

We have known tea-drinkers, too, who after a hard day's toil, could drink two or three cups of strong tea and lie down to sleep for the night as quietly as babes are expected to — but do not.

It must be evident that each person should observe the effects of tea upon himself or herself and be governed accordingly. Tea is a poison to some temperaments, and so are strawberries. Tea will cure a headache or may produce one; will dispose to rest or excite to action. We will sum then by conceding that all our quoted authorities are right in their conclusions, if limited to a limited class of tea-drinkers, and all are wrong, in a very broad application.

Theine is the one constant agency in the effects of tea. It is present in teas that are devoid of essential oils — so far as the senses go — and it then still refreshes, stimulates, sustains, and even exhilarates, by actual experiment.

The feeling of "comfort," attributed by some writers to the hot water of the tea, may be also enjoyed by drinking cold tea, which is no less refreshing in hot weather. The high-flavored essential oils (strictly oils which evaporate at very moderate temperatures) of Formosa teas seem to take part in the superior exhilarating or almost intoxicating effects of the choice varieties, but we have no certain proof of the fact; while the more intoxicating and stimulating, as well as deleterious, green teas possess very little, if any, of these pleasant oils.

It seems to be an authodox opinion among physiologists that tea contributes nothing towards support of the human system; that it only rouses it into action, an effect which should, consistently, be followed by corresponding reaction and depression, which plainly is not the case. This hypothesis leaves the enquiring layman in a dilemma. Tea must either enable the system to draw more heavily or more economically upon the resources afforded by recognized food, or it is itself nutriment. Otherwise, an established principle of physics — that there can be no expenditure of energy without correlative cost — would be subverted. As tea is admitted upon experience to be most useful, and most craved by mankind, where the supply of food is insufficient; and as it is known to refresh and sustain in large degree in the absence of any food whatever, there is fair ground for the opinion, however heterodox, that tea directly affords nutriment to the human organism, and, possibly, to the brain and nerves in particular, as with phosphoric acid.

Animal gelatine has been placed in the same class with tea by Liebig, Dr. John W. Draper, and others, and it is asserted that it conserves waste without itself entering into the substance of human tissue. It is an accepted physiological law that nothing taken as food or drink can support expenditure of human energy in sensible motion, in heat, or in the nervous waste of mental or emotional exercise without first being built up into living tissue; the breaking down or chemical decomposition of which tissue, and subsequent oxidation of less complex compounds or their constituents, is the direct source of bodily energy of every description. This, at least, is our reading of modern authorities, like Foster. If tea and gelatine, and possibly alcohol, are to form exceptions to the law, the law no longer stands. But it would seem more reasonable to amend the hypothesis concerning exceptions, and bring them into line by admitting that they are nutritious in a manner not yet ascertained. All physiological laws are provisional, good until proved insufficient, and then to be amended in the light of accumulating facts.

CHAPTER VIII.

Meanwhile Hanna the housemaid had closed and fastened the shutters, Spread the cloth, and lighted the lamp on the table, and placed there Plates and cups from the dresser, the brown rye loaf and the butter Fresh from the dairy, and then, protecting her hand with a holder, Took from the crane in the chimney the steaming and simmering kettle, Poised it aloft in the air, and filled the earthen teapot, Made in delft, and adorned with quaint and wonderful figures.

LONGFELLOW'S TALES OF A WAYSIDE INN.

Many besides those who live principally by the labor of their brains, will subscribe to the sentiment expressed by Thomas De Quincey, in his Confessions of an English Opium Eater, when he said that—"Tea, though ridiculed by those who are naturally of coarse nerves, or are become so from wine drinking, and are not susceptible of influence from so refined a stimulant, will always be the favorite beverage of the intellectual; and for my part, I would have joined Dr. Johnson in a bellum internecinum against Jonas Hanway, or any other impious person who should presume to disparage it."

The only stimulant that Hazlitt indulged in was strong Black tea, using the very best obtainable.

Wordsworth was a lover of tea, and he sweetened his tea beyond the taste of ordinary mortals.

Shelly also was a lover of tea. Kant drank tea habitually for breakfast. Motley used either tea or coffee for breakfast, as fancy prompted.

William Howitt found great refreshment in both tea and coffee, but he wrote that on his great pedestrian journeys, "Tea would always in a manner almost miraculous banish all my fatigue, and diffuse through my whole frame comfort and exhilaration without any subsequent evil effect. Tea is a wonderful refresher and reviver."

Justin McCarthy, M. P. the brilliant historian, said that he was a liberal drinker of tea, and that he found it "of immense benefit in keeping off headache, my only malady."

Harriet Martineau dearly loved her cup of tea. Geo. R. Sims says "Tea is my favorite tonic when I am tired or languid."

An amiable weakness for Afternoon Tea in the course of his daily official duties which was manifested by the late Hon. Wm. L.

Strong, the worthy mayor of New York in 1895-6, furnished the New York newspapers with opportunities for many a good-natured jest and jibe; one of the best of which we have preserved in the lines which follow.

A BALLAD OF OOLONG.

By John Paul Bocock.

Whenever the magistrate, good Li Song
Is short of his favorite tea, Oolong,
He lays his gout and his spectacles down
And hies him away into Chinatown.

Into the region of Mon Lay Won,
When the day of official life is done,
Into the land of slant-eyed Lee's
He hies him away to replenish his teas.

All day long, in the places of Tax,
Of rubicund tape and sealingwax,
He toils and moils till the hour of tea,
Blessed old five o'clock, sets him free!

Blest liberator, better than rum,
Of the Fa and the Fee and the Fi Fo Fum
Of the tammany Ogre who used to dwell
In the metropolitan citadel.

Blest over all the heroes that be
On the sunny side of the Ceylon Sea,
Nerve him still to be Good and Strong.
Excellent magistrate, great Li Song.

Dr. King Chambers, in a Manual of Diet in Health and Disease says of Tea that—"It soothes the nervous system when it is in an uncomfortable state from hunger, fatigue, or mental excitement."

Florence Nightingale said—"When you see the natural and almost universal craving in English sick for their tea, you cannot but feel that nature knows what she is about. There is nothing yet discovered which is a substitute to the English patient for his or her cup of tea."

Buckle (the Historian) quotes Dr. Jackson as saying (in 1845) that—"Even for those who have to go through great fatigues, a breakfast of tea and dry bread is more strengthening than one of beefsteak and porter."

Prof. Parkes says—"As an article of diet for soldiers, tea is most useful. The hot infusion, like that of coffee, is potent both against heat and cold; it is useful in great fatigue, especially in hot climates, and also has a great purifying effect upon water. It should form the drink par excellence of the soldiers on service."

Admiral Inglefield, in 1881, said, that in evidence given before the Artic Committee, of which he was a member, all the witnesses were unanimous in the opinion that spirits taken to keep out cold was a fallacy, and that nothing was more effectual than a good fatty diet, and hot tea or coffee, as a drink "Seamen who Journeyed with me up the shores of Wellington Channel," says the Admiral," in the artic regions, after one day's experience of rum-drinking, came to the conclusion that Tea, which was the only beverage I used, was much more to be preferred."

Lord Wolsely, late Commander in Chief of the British Army, wrote as follows:—

"It fell to my lot to lead a brigade through a distant country for more than 600 miles. I fed the men as well as I could, but no one, officer or private, had anything stronger than tea to drink during the expedition. The men had peculiarly hard work to do, and they did it well, and without a murmur. We seem to have left crime and sickness behind us with the 'grog,' for the conduct of all was most exemplary and no one was ever ill. "

Mr. Winter Blyth, Medical Officer of Health for Marylebone, (London), says in reference to long cycling excursions, and experiments with beer and spirits,—"My own experience as to the best drink when on the road is most decidedly in favor of Tea. Tea appears to rouse both the nervous and muscular systems, with, so far as I can discover, no after-depressing effects."

"Edward Payson Weston, the great Pedestrian, finds in Tea and rest the most effective restoratives. He once walked 5000 miles in 100 days, and after each day's work, lectured on 'Tea versus Beer.'"

C. J. Nichod, late Secretary of the London Athletic Club, writes in his book—"Guide to Athletic Training," that "Tea is preferable for training purposes, possessing less heating properties and being more digestible than beer or spirits."

Cowper's lines, however hackneyed in quotation, are still classic in their application to English homes and their evening accompaniment, Tea.

"Now stir the fire, and close the shutters fast,
Let fall the curtains, wheel the sofa round,
And while the bubbling and loud-hissing urn
Throws up a steamy column, and the cups
That cheer but not inebriate, wait on each,
So let us welcome peaceful evening in."

"Tea" was the designation of the customary evening meal in most American families for about two centuries, and as late as 1850, since which time it has merged in the more substantial "late dinner," in cities and towns especially, although the last meal of the closing day is still "Tea" in spirit and in name in many families where commercial necessities have not compelled change. The same is true of England from which we derive our customs, and with which we also changed it. According to Washington Irving's veracious History of New York, tea-parties were indulged in by the Dutch inhabitants of New Amsterdam during the reign of Governor Wouter Van Twiller (which commenced in 1633). Irving says:

"But though our worthy ancestors were singularly averse to giving dinners, yet they kept up the social bonds of intimacy by occasional banqueting, called tea parties.

"These fashionable parties were generally confined to the higher classes or noblesse, that is to say, such as kept their own cows, and drove their own wagons. The company commonly assembled at 3 o'clock, and went away about six, unless it was in winter time, when the fashionable hours were a little earlier, that the ladies might get home before dark. . . . The tea was served out of a majestic Delft tea-pot, ornamented with paintings of fat little Dutch shepherdesses tending pigs, with boats sailing in the air and houses built in the clouds, and sundry other Dutch fantasies. The beaux distinguished themselves by their adroitness in replenishing this tea-pot from a huge copper tea-kettle. . . .

To sweeten the beverage, a lump of sugar was laid beside each cup, and the company alternately nibbled and sipped with great decorum, until an improvement was introduced by a shrewd and economic old lady, which was to suspend a large lump directly over the tea-table by a string from the ceiling, so that it should be swung from mouth to mouth—an ingenious expedient which is still kept up by some families in Albany, but which prevails without exception in Communipaw, Bergen, Flatbush, and all our uncontaminated Dutch villages.

"At these primitive tea-parties the utmost propriety and dignity of deportment prevailed. No flirting or coquetin gambu of old ladies, nor hoyden chattering and romping of young ones, no self satisfied struttings of wealthy gentlemen with their brains in their pockets, nor amusing conceits and monkey divertisements of smart young gentlemen with no brains at all. On the contrary, the young ladies seated themselves demurely in their rush-bottomed chairs, and knit their own woolen stockings, nor ever opened their lips except to say "yaw, mynherr," or "yaw, yaw, Vrouw," to any question that was asked them, behaving in all things like decent, well educated damsels. As to the gentlemen, each of them tranquilly smoked his pipe, and seemed lost in contemplation of the blue and white tiles with which the fire-places were decorated, wherein sundry passages of scripture were piously portrayed."

But it was in New England that the tea-party reached its highest importance as a social function, and in the New England of more than a century ago. Then and there were the weightiest themes of religion and philosophy of such enthralling interest and so interwoven with the practical affairs of men, that they were familiarly discussed all the way from the pulpit and desk to the household and tea-table, and were liable to be brought forward at the table of the artisan, the farmer, or the shopkeeper, as well as at that of the scholar. Every reader of early New England history or New England fiction must be aware of this fact. The presence of the "minister." so far from discouraging these discussions, usually stimulated them, and lent them additional interest. Instances of such gatherings and conversations, of typical New England tea-parties, may be found in Mrs. Stowe's Minister's Wooing.

The "tea-table" will always live in name and in association, and we trust in reality, as an essential feature of family life, even though the nature of the repast has greatly changed. The pleasantest part of the working-day in former years was the occasion when the family, drawn together by common interests and sympathies, after the heavier tasks of the day were completed, gathered around the table whose crowning symbol of good cheer was the familiar and homely old tea-pot. From this fairy godmother flowed forth a spirit of kindly toleration and genial good humor.

A quiet fireside, a snug corner, and a singing tea-kettle, were potent sources of enjoyment to young as well as old folks, in those days when the kitchen was not turned entirely over to alien hands.

The tea-kettle and the hearth-stone may be pushed back out of sight or even quite banished from the household, by modern metropolitan life and enforced changes; but under the influence of old associations and traditions, they will surely return in time with recurring cycles of sentiment or of fashion.

Five o'clock Tea is but an attempt to revive an old custom, and for those whom fortune has favored with leisure for social amenities at that hour, it furnishes an agreeable and informal occasion for exchange of courtesies and for harmless gossip or even more dignified "conservation."

A correspondent of the New York Sun recently gave an account of actual or impending changes in the social customs of Paris, which have a bearing upon this branch of our subject. He writes that the English five o'clock tea having been adopted by Parisians several years ago, and being found to interfere with the still fashionable 7 o'clock dinner, an effort was recently made to revive the ancient mid-day dinner, say at 2 o'clock. In some cases, the difficulty was met by taking tea at five o'clock, and serving a substantial supper late in the evening.

When we desire to get away for a time from our modern conventional ideas and restraints, and indulge in a bit of homely healthy sentiment, we may fall back on such utterances as the following, from Dicken's Cricket on the Hearth:

"Now it was, you observe, that the Kettle began to spend the evening. Now it was, that the Kettle, growing mellow and musical, began to have irrepressible gurglings in its throat, and to indulge in short vocal snorts, which it checked in the bud, as if it hadn't quite made up its mind yet, to be good company. Now it was, that after two or three such vain attempts to stifle its convivial sentiments, it threw off all moroseness, all reserve, and burst into a stream of song so cosy and hilarious as never maudlin nightingale yet formed the least idea of." . . .

"So plain, too! Bless you, you might have understood it like a book—better than some books you and I could name, perhaps. With its warm breath gushing forth in a light cloud which merrily and gracefully ascended a few feet, then hung about the chimney-corner as its own domestic Heaven, it trolled its song with that strong energy of cheerfulness, that its iron body hummed and stirred upon the fire, and the lid itself, the recently rebellious lid—such is the influence of a bright example— performed a sort of jig, and clattered like a deaf and dumb young cymbal that had never known the use of its twin brother." . . .

"And here, if you like, the Cricket DID chime in! with a Chirrup, Chirrup, Chirrup of such magnitude, by the way of chorus, with a voice so astoundingly disproportionate to its size, as compared with the Kettle, (size! you couldn't see it!) that if it had then and there burst itself like an overcharged gun, if it had fallen a victim on the spot, and chirruped its little body into fifty pieces it would have seemed a natural and inevitable consequence for which it had expressly labored." . . .

"There was all the excitement of a race about it. Chirp, chirp, chirp! Cricket a mile ahead. Hum, hum, hum-m-m! Kettle making play in the distance, like a great top. Chirp, chirp, chirp!—Cricket round the corner. Hum, hum, hum! Kettle sticking to him in his own way, no idea of giving in. Chirp, chirp, chirp ! Cricket fresher than ever. Hum, hum, hum-m-m! Kettle slow and steady. Chirp, chirp, chirp! Cricket going to finish him. Hum, hum, hum! Kettle not to be finished. Until at last, they got so jumbled up together, in the hurry-skurry, helter-skelter of the match, that whether the Kettle chirped or the Cricket hummed, or the Cricket chirped and the Kettle hummed, or the Cricket chirped and the Kettle hummed, or the both chirped and both hummed, it would have taken a clearer head than yours or mine to have decided with anything like certainty. But of this there is no doubt, that the Kettle and the Cricket, at one and the same moment, and by some power of amalgamation best known to themselves, sent each his fireside song of comfort streaming into a ray of the candle that shone through the window, and a long way down the lane. And this light, bursting on a certain person who on the instant, approached towards it through the gloom, expressed the whole thing to him, literally in a twinkling, and cried, 'Welcome home, old fellow! Welcome home, my boy!"

CHAPTER IX.

"The willow-pattern that we knew
In childhood. with its bridge of blue,
Leading to unknown thoroughfares."
— —Keramos, Longfellow.

Peradventure some who read these rambling paragraphs may be the fortunate possessor of a few pieces of that willow-pattern, blue or pink china table ware which was but too lightly esteemed when it was a common heritage of English and American families. If not, a vivid remembrance of the ware and of the fancies which it inspired, must be little less prized by those who cherish such associations with home and childhood. We are tempted here to recall some of our own reminiscences of old china, which the impatient reader may excusably skip for more serious matter.

From the semi-aquatic summer-house with roof curving upward like an inverted umbrella, imprinted upon a favorite tea-plate, we often sallied forth in fancy to explore the Chinese world as portrayed in blue or pink upon earthen table-ware of the olden time. And what a world! How artfully adapted to childish notions, how convenient for bird's-eye views, this arrangement of lofty mountain peaks, deep gorges, and rocks of fantastic forms, tangled up with examples of nature subdued by Chinese art in landscape gardening and ornate architecture. In the near distance (far and near are the same in Chinese art), we behold a slender streak of waterfall descending from mountain peaks a thousand feet or height by comparison; a broad flight of stone stairs leading up to a palace or temple of intricate construction and marvellous ornamentation; a majestic river a mile or two in width, winding serenely by these wonders of nature and art, but submitting to be spanned by a single arch of bridge, perhaps thrice the length of the Chinaman advancing over its camel-humped back, who placidly regards from under his ruffle-edged umbrella the pleasure boats floating beneath him. A little group of high- born Chinese ladies in holiday attire are seated in a garden of potted plants on the river's bank, drinking tea, flirting their fans, and doubtless talking over the latest Court gossip. Nearby is a willow, not the stiff, ugly tree now seen upon tame and degenerate imitations of real old China pottery, but a graceful weeping-willow, whose drooping branches sweep the opposite shore, as sublimely indifferent to distance as the untrammeled artist himself.

No hint here of imperative human toil, or of human need, or of anything but present enjoyment and rest; it is a picture of contented, comfortable existence, for dreamy contemplation, amid a grouping of art and nature that calmly defies probability and challenges the impossible.

But perhaps the Chinese artist had more justification for his incredible fancies than we have imagined. Strange contradictions occur in China, judged by our conventional standards, and there are surprises and incongruities even in their actual landscapes, which are unsuspected by thousands of our intelligent countrymen. Some examples of such departure from our notions of natural and of artificial scenery are given in the illustrations of this work.

CHAPTER X.

"The east wind fans a gentle breeze,
The streams and trees glory in the brightness of the spring.
The bright sun illuminates the green shrubs,
And the falling flowers are scattered and fly away,
The solitary cloud retreats to the hollow hill;
The birds return to their leafy haunts:
Every being has a refuge whither he may turn;
I alone have nothing to which to cling.
So, seated opposite the moon shining o'er the cliff,
I drink and sing to the fragrant blossoms."

The foregoing lines are by Le Tai-Pih, styled the Chinese Anacreon, literally translated by R. K. Douglas, in the Encylopaedia Britannica. They might easily apply to a tea garden.

The power of a single word to arouse trains of thought composed of the most varied ideas, to set in motion a panorama of scenery which is well nigh endless with persons of lively imaginations, is illustrated by this word, tea. While to one person it may suggest only refreshment and personal comfort, and to another, scenes of home life, to still others it will bring into being all that the dreamer has read or heard of China, that land of Cathay, and of its slant-eyed, mild mannered wearers of the pig-tail, and their real or fabulous characteristics. Not the least interesting of such associations are memories of the queer manners and habits of the Chinese people, some of which to us outside barbarians, appear so drolly opposed to our civilization of fancied superiority. Let us recapitulate a few of the most marked differences between the Chinese and Western peoples.

The very first antithesis that strikes us is the braided pig-tail of long black glossy hair so religiously cherished by the men. Have they forgotten that this is a badge of servitude? The original inhabitants of China—by which we mean that people who occupied central China as far back as the beginning of the Assyrian Empire, or say 1300 years before Christ,—are said to have worn their jet black hair long, and coiled loosely upon the crown of the head, but they did not shave any portion of the head, nor braid their hair in a queue. The northern tribes of Manchus and Mongols (Tarters or Taters in olden nomenclature), who inhabited Manchuria and Mongolia, had endeavored to conquer the Chinese in wars which began about 950 A. D., and during which in the 12th century, the celebrated Jenghiz Khan and Kublai Khan severally commanded the Mongolian armies. These wars continued until 1627 A. D. when the Manchurian invaders regarded their conquest as sufficiently assured to warrant them in imposing their commands upon their Chinese vassals. At that time the Manchus partly shaved their head and wore braided queues. In 1627 an edict was issued by the Manchus requiring all Chinese subjects to henceforth follow the Manchu fashion and to wear the pig-tail as a token of submission to their conquerors. So, after time a badge of bondage became with the Chinese an insignia of national pride and honor.

Then, let us consider their written language, the oldest in the world except Hebrew, says Dr. Williams, and the oldest spoken language without any exception. Professor James Legge, writing upon Confucianism and Taoism, says that the written language of China takes us back at least five thousand years. Like most things in China, the language has suffered very little change since its adoption and completion. It does not consist of words, built up of letters, as with us; it has no alphabet, no letters, but its curious symbols represent objects, qualities, ideas, or sounds, which by combination express every shade of Chinese thought. The number of these written characters is variously estimated by European philologists at from 25,000 to 50,000, although it is believed that one may become a fair reader of Chinese literature, by acquiring a knowledge of say 10,000 of the pictorial symbols, with their allowable variations of form in use. Punctuation is not ordinarily used in Chinese literature and of course sentences or paragraphs are not divided from each other by capitals, for they have none.

In the spoken language, rising or falling inflections, and indescribable variations of tone must be learned, as well as pronunciation, and when it is said that there are many different dialects, each unintelligible to those accustomed to some other one, there seems to be little encouragement for the introduction of Chinese into our public school system. For all this, Dr. Morrison, the compiler of a Chinese and English dictionary, declares that "Chinese fine writing darts upon the mind with a vivid flash, a force and beauty, of which alphabetic language is not capable."

Graphic representation of an idea in a picture illustrates Dr. Morrison's meaning.

Chinese written or printed composition is arranged in perpendicular columns, which are read from top to bottom and from the right to the left; and a Chinese book begins at the end from our point of view.

When in China two polite acquaintances accost each other, they pause before meeting and each shakes his own hand; (a much neater and more refined custom than our own).

To raise one's hat to a Chinaman is to offer an insult.

A favorite road vehicle for passengers is a wheel barrow, and a mast and sail are often attached to aid in its propulsion, with a fair wind.

Kite-flying is a sport for old men, boys look on.

The game of checkers or draughts is played with 360 men.

Shop signs are set on end.

White is the universal color for full mourning. Men make women's head dresses.

Women row heavy boats on the canals.

A Chinese compass needle points to the south.

In addressing a person, his last or surname is first written, and his first name last.

The seat of honor at the table is at the left of the host.

Fashions in fine clothing never change in China.

> Thieves are required by the Government to be organized into companies or guilds with elected heads, with whom the Government and public may treat.

> If a man is busy at his store, a traveling restaurant will wait upon him.

> A charcoal furnace, culinary vessels, and food, are slung upon a pole carried by the proprietor, who stops before the customer's door, and cooks a meal to order.

> The first paddle-wheel boats built in China were anchored in the stream where the current turned the paddle-wheels, and ground grain for food.

The Chinese paint the edges of their shoe-soles white.

> An expensive coffin is always an acceptable present from an affectionate son and heir to his living father.

> Military officers in the Chinese army formerly wore embroidered silk petticoats, and strings of beads around their necks; they carried fans, and mounted their horses on the right hand side.

Chinese Cashiers are said to be uniformly honest.

CHAPTER XI.

American Tea Culture.

During a period of at least 40 years, tea plants have been cultivated by a few experimenters in the southern United states, and American tea, grown South Carolina, Georgia, and Florida, has satisfactorily supplied the family needs of a hundred or more persons, at a cost not exceeding the retail price of good foreign tea.

When Mr. Wm. G. Le Duc, Commissioner of the Department of Agriculture at Washington, seriously recommended systematic tea culture in the southern States, press writers and press readers found a new subject of mirth and standing jokes which lasted for several years. To be sure, those who laughed so long and loudly did not know the difference between a Chinese tea plant and a China Aster, and few of them had ever heard that in certain tea growing districts of China, ice and snow were familiar associates of the hardy Chinese tea plant. Enquiry would have taught them that here in the United States individual tea plants had for many years withstood a freezing temperature in winter. Better informed persons fell back upon the objection that Americans could never learn the secrets of curing tea, and finally that the very low cost of Chinese labor would be fatal to American competition. But the mills of the Gods grind right along, regardless of individual opinions or precedents. Foreign tea plants have been so acclimatised in South Carolina that a plantation of tea has withstood a winter temperature of zero, the lowest recorded degree for 150 years; the secrets of curing the leaf have been disclosed and successfully practiced by Americans, and a cheap form of child labor for picking the tea leaves has resulted in commercial success for American grown tea.

This result is due to the encouragement of the U. S. Agricultural Bureau, and the persistent efforts of Dr. Charles U. Shepard, at Summerville, S. C., who continued his exertions to found a permanent tea plantation on a large scale long after the Government authorities had ceased to hope for success. In Dr. Shepard's tea gardens the deficiency in rain fall is made good by deep pulverization of the soil and artificial irrigation; the natural shade of jungle or forest under which the seed germinates and grows where the plant is indigenous, is supplied by artificial shade; and the expensive process of picking the leaves is cheapened by employing children, who are paid in money, and also by being taught to read and write in a school maintained on the premises by Dr. Shepard. Machinery has supplanted some of the tedious hand-manipulation of tea in Dr. Shepard's factory, and further progress in this direction is constantly being made.

The Pinehurst tea—for Pinehurst is the designation of Dr. Shepard's plantation at Summerville—sometimes disappoints those accustomed to the strong flavors and pronounced fragrance of some foreign teas, but it contains a full proportion of that stimulating, sustaining constituent of all genuine teas, theine, as consumers all discover. Like our American grapes and wines, American teas will doubtless improve by continuous cultivation upon a given soil, and probably will at length develop characteristics of their own, as precious in the estimation of tea drinkers as those of the exceptional foreign teas.

Impressed by the importance of Dr. Shepard's success, and the latent possibilities of this new field of American enterprise, Messrs. Francis H. Leggett & Co., of New York, have purchased from Dr. Shepard the entire crop of American Pinehurst teas for 1900, amounting in quantity to several thousand pounds.

CHAPTER XII.

How Shall We Make Tea?

How shall tea be drawn or infused? Is there but one standard method for all teas, or all persons? Certainly not. A method which will suit very many delicate tastes may be briefly stated: Use water as free as possible from impurities, from earthly matters like lime. If water is boiled too long its contained air is expelled and the tea will have a "flat" taste. Use an earthen teapot by preference; one which is never applied to any other purpose. A preliminary warming of the dry teapot is advised. Drop in your tea leaves, and pour on the whole quantity of water required, while at boiling temperature. Set in warm but not very hot situation to steep, avoiding so far as practicable, loss of vapor and aroma from the teapot.

Now, as to the length of time tea should steep: — it will vary with different teas and different tastes. Some steep tea but three minutes; others double the time; while still others extend the time to 15 minutes. In any event, as soon as the characteristic flavor is extracted from the leaves, known by the loss of an agreeable tea-odor in the withdrawn leaves, the beverage will be improved rather than impaired by pouring it off into a clean teapot, in which the tea may then be preserved for a long time without injury.

To some tastes, a little of the tannin is agreeable, and its absence would be missed. Then as to sugar or milk: it is evidence of exaggerated personality (conceit, some call it), to declare that milk or cream or sugar injure the flavor of tea. As well insist upon a special spice being used for all viands because the critic likes it. To hold the Chinese up as examples of what is proper in tea drinking is to offer a limit to human progress. As milk or cream neutralize the tannin to a considerable extent, they are so far desirable, without regard to taste.

OVER MY TEA CUP.

by Charles J. Everett

This homely can of painted tin
Is casket precious in my eyes;
Its withered fragrant leaves within,
Beyond all costly gems I prize.
For for those crumpled leaves of tea,
The sunbeams of long summer days,
The song of bird, the hum of bee,
The cricket's evening hymn of praise,
The gorgeous colors of sunrise,
The joy that greets each new-born day;
The glowing tints of sunset's skies,
The calm that comes with evening grey;
The chatter of contented toil,
The merry laugh of childish glee,
The tonic virtues of the soil,
Were caught and gathered with the tea.
Lifeless those withered leaves may seem,
Locked fast in slumber deep as death,
But soon the Kettle's boiling steam
May rouse to life their fragrant breath.
With sigh of deep content we breath
The sweet mists rising lazily,
With eager, parted lips receive taste of tea.
Forlight and warmth and mood of men,
Whate'er the plant hath heard or seen
Or felt, while fixed in field or fen,
And stored within its depths serene,
Are now transmuted into thrills
Of sense or feeling, echoes faint
From peaceful perfumed tea-cladhills,
From placid Orientals quaint.
And fancies born in other lands,
Which dormant lie in magic tea,

Dream-castles fair not made with hands,
By some mysterious alchemy
Emerge from cloudland into sight,
Transform the sombre working-world,
The gloomy hours of day or night
From leaden hue to tint of gold,
Bring rest to wearied heart and brain,
Kind nature's soul to us reveal,
Enlarge the realm of Fancy's reign,
Renew the power to see and feel
The radiance of the rising sun,
The sunset's glow, the moon's pale light,
The promise of a day begun,
The rest from toil that comes with night.
And as I sip my cup of tea,
Though not a friend may be in sight,
I know that a brave company
Is taking tea with me this night.

www.ingramcontent.com/pod-product-compliance
Lightning Source LLC
Chambersburg PA
CBHW030456220526
45464CB00006B/2559